ACHIEVE

The higher
score

Year
6

Mathematics

SATs Revision

Trevor Dixon
& Sarah-Anne Fernandes

RISING STARS

Orders: Please contact Hachette UK Distribution, Hely Hutchinson Centre, Milton Road, Didcot, Oxfordshire, OX11 7HH. Telephone: (44) 01235 400555. Email: primary@hachette.co.uk.

ISBN: 978 1 51044 270 2

© Hodder & Stoughton Ltd (for its Rising Stars imprint) 2019

This edition published in 2018 by Rising Stars, part of Hodder & Stoughton Ltd.
First published in 2015 by Rising Stars, part of Hodder & Stoughton Ltd.
Rising Stars UK is part of the Hodder Education Group
An Hachette UK Company
Carmelite House
50 Victoria Embankment
London EC4Y 0DZ

www.risingstars-uk.com

Impression number   10 9 8 7
Year 2023

Authors: Trevor Dixon and Sarah-Anne Fernandes

Educational Adviser: Steph King

Series Editor: Sarah-Anne Fernandes

Accessibility Reviewer: Vivien Kilburn

Cover design: Burville-Riley Partnership

Illustrations by Ann Paganuzzi

Typeset in India

Printed in India

A catalogue record for this title is available from the British Library.

Contents

Welcome to Achieve Mathematics: The Higher Score – Revision

In this book you will find lots of practice and information to help you achieve the higher score in the Key Stage 2 Mathematics tests. You will look again at some of the same key knowledge that was in Achieve Mathematics: The Expected Standard, but you will use it to tackle trickier questions and apply it in more complex ways.

About the Key Stage 2 Mathematics National Tests

The tests will take place in the summer term in Year 6. They will be done in your school and will be marked by examiners – not by your teacher.

There are three papers to the tests:

Paper 1: Arithmetic – 30 minutes (40 marks)
* These questions assess confidence with a range of mathematical operations.
* Most questions are worth 1 mark. However, 2 marks will be available for long multiplication and long division questions.
* It is important to show your working – this may gain you a mark in questions worth 2 marks, even if you get the answer wrong.

Papers 2 and 3: Reasoning – 40 minutes (35 marks) per paper
* These questions test mathematical fluency, solving mathematical problems and mathematical reasoning.
* Most questions are worth 1 or 2 marks. However, there may be one question with 3 marks.
* There will be a mixture of question types, including multiple-choice, true/false or yes/no questions, matching questions, short responses such as completing a chart or table or drawing a shape, or longer responses where you need to explain your answer.
* In questions that have a method box it is important to show your method – this may gain you a mark, even if you get the answer wrong.

You will be allowed to use: a pencil/black pen, an eraser, a ruler, an angle measurer/protractor and a mirror. **You are not allowed** to use a calculator in any of the test papers.

Test techniques

Before the tests
- Try to revise little and often, rather than in long sessions.
- Choose a time of day when you are not tired or hungry.
- Choose somewhere quiet so you can focus.
- Revise with a friend. You can encourage and learn from each other.
- Read the 'Top tips' throughout this book to remind you of important points in answering test questions.
- Make sure that you know what the bold key words mean.
- For longer questions, remember to practise showing your working: this could help you gain extra marks in the tests.

During the tests
- READ THE QUESTION AND READ IT AGAIN.
- If you find a question difficult to answer, move on; you can always come back to it later.
- Always answer a multiple-choice question. If you really can't work out an answer, try to think of the most sensible response and read the question again.
- Check to see how many marks a question is worth. Have you written enough to 'earn' those marks in your answer?
- Read the question again after you have answered it. Make sure you have given the correct number of answers within a question, e.g. if there are two boxes for two missing numbers.
- If you have any time left at the end, go back to the questions you have missed.

Where to get help
- **Pages 8–9** practise number and place value.
- **Pages 10–19** practise number – addition, subtraction, multiplication and division.
- **Pages 20–28** practise number – fractions, decimals and percentages.
- **Pages 29–32** practise ratio and proportion.
- **Pages 33–36** practise algebra.
- **Pages 37–45** practise measurement.
- **Pages 46–47** practise geometry – properties of shapes.
- **Pages 48–53** practise geometry – position and direction.
- **Pages 54–61** practise statistics.
- **Pages 62–63** provide definitions of all the key words.
- **Pages 67–68** provide the answers to the 'Try this' questions.
- **Inside back cover** contains a revision checklist to help you keep track of your progress.

How to use this book

1 ***Introduction*** – each content strand in the mathematics National Curriculum has been broken down into smaller topics. This introduction tells you what you need to be able to do for this topic.

2 ***What you need to know*** – summarises the key information for the topic. Words in bold are key words and those in lilac are also defined in the glossary at the back of the book.

3 ***Let's practise*** – a practice question is broken down in a step-by-step way to help you to understand how to approach answering a question and get the best marks that you can.

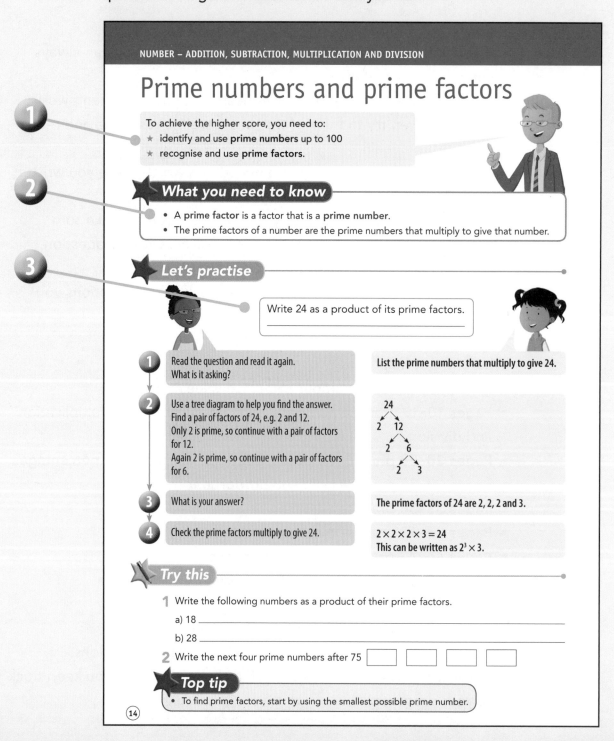

NUMBER – ADDITION, SUBTRACTION, MULTIPLICATION AND DIVISION

Prime numbers and prime factors

1

To achieve the higher score, you need to:
* ★ identify and use **prime numbers** up to 100
* ★ recognise and use **prime factors**.

2

What you need to know

* A **prime factor** is a factor that is a **prime number**.
* The prime factors of a number are the prime numbers that multiply to give that number.

3

Let's practise

Write 24 as a product of its prime factors.

1	Read the question and read it again. What is it asking?	List the prime numbers that multiply to give 24.
2	Use a tree diagram to help you find the answer. Find a pair of factors of 24, e.g. 2 and 12. Only 2 is prime, so continue with a pair of factors for 12. Again 2 is prime, so continue with a pair of factors for 6.	24 2 12 2 6 2 3
3	What is your answer?	The prime factors of 24 are 2, 2, 2 and 3.
4	Check the prime factors multiply to give 24.	$2 \times 2 \times 2 \times 3 = 24$ This can be written as $2^3 \times 3$.

Try this

1 Write the following numbers as a product of their prime factors.

  a) 18 _____

  b) 28 _____

2 Write the next four prime numbers after 75 ☐ ☐ ☐ ☐

Top tip

* To find prime factors, start by using the smallest possible prime number.

14

4 *Try this* – this is where you get the chance to practise answering questions for yourself. There are a different number of questions for each topic.

5 *Top tips* – these give you further reminders about answering test questions or help you to understand a tricky topic.

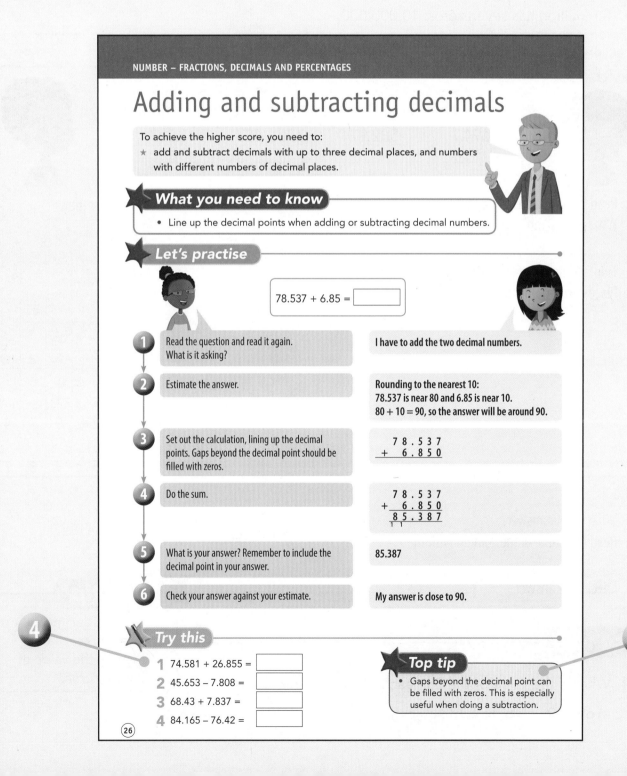

Place value

To achieve the higher score, you need to:
★ know the **place value** of numbers up to 10,000,000 with up to three **decimal places**.

What you need to know

- The **place value** of each digit in a number depends on its position.
- One million has six zeros: 1,000,000.
- Ten million has seven zeros: 10,000,000.

- Numbers following the decimal point are fractions, showing tenths, hundredths and thousandths.

Let's practise

What is the value of the 7 in each of these numbers?

694.7   582.974   9,706,453   45.967

_____   _____   _____   _____

1 Read the question and read it again. What is it asking?

Identify the value of 7 in each number.

2 Think of the value of the columns. Place each number in the place value table.

Millions	Hundreds of thousands	Tens of thousands	Thousands	Hundreds	Tens	Ones	.	tenths	hundredths	thousandths
M	Hth	Tth	Th	H	T	O	.	t	h	th
				6	9	4	.	7		
				5	8	2	.	9	7	4
9	7	0	6	4	5	3	.			
				4	5	.	9	6	7

3 What is your answer? Read the name of the column with the 7.

$\frac{7}{10}$ $\frac{7}{100}$ 700,000 $\frac{7}{1,000}$

4 Check your answer.

Top tip

- Use the decimal point as a marker to help identify the value of each column.

Try this

1 What is the value of the **3** in each of these numbers?

a) 698.39   b) 894.738   c) 5.043   d) 7,093,528 e) 5,632.879

_____   _____   _____   _____   _____

Roman numerals

To achieve the higher score, you need to:

★ read **Roman numerals** to 1,000 (using I, V, X, L, C, D and M).

What you need to know

- With **Roman numerals**, letters are used to represent numbers.

  I = 1   C = 100
  V = 5   D = 500
  X = 10   M = 1,000
  L = 50

- The letters are arranged one after another and their values added (e.g. the number 72 is split into 50 + 10 + 10 + 1 + 1 and written as LXXII).

- If a letter is placed before another letter of greater value, subtract that amount (e.g. IV = 4 [5 – 1 = 4] XC = 90 [100 – 10 = 90] CM = 900 [1,000 – 100 = 900]).

Let's practise

Write CCCLIX in figures. ⬚

1 Read the question and read it again. What is it asking?

Turn CCCLIX into a number.

2 Be systematic. Convert one letter at a time.

C means 100 so CCC means 300.
L means 50 so CCCL means 350.

3 Watch out for letters that are out of numerical sequence. I < X so you have to subtract.

IX means 1 less than 10.

4 What is your answer?

CCCLIX means 359.

5 Check your answer by converting it back.

100 + 100 + 100 + 50 + 9 = 359

Try this

1 Write these numbers in Roman numerals.

a) 265 ⬚   b) 999 ⬚   c) 2,000 ⬚   d) 2,015 ⬚

2 Convert these Roman numerals into decimal figures.

a) MMXIV ⬚   b) CCLXXXVI ⬚

c) LXXXIX ⬚   d) CXXIV ⬚

Top tip

- Convert each letter to its value. Letters in front of a letter of a higher value are subtracted. Then you can add the values together.

Addition and subtraction

To achieve the higher score, you need to:
★ solve multi-step addition and subtraction problems, deciding which operations and methods to use and why.

What you need to know

- Key words in a problem help you to make sense of the operations that have been used or that you will need to use:
  - The words *total of*, *increase by*, *plus* or *altogether* relate to addition.
  - The words *difference between*, *reduce by*, *minus* or *less* relate to subtraction.
  - Think carefully about *more than* and *less than* questions as the operation will depend on the problem.

Let's practise

$$5,632 + \boxed{} + 329 = 9,802$$

1 Read the question and read it again. What is it asking?

Find the missing number in an addition calculation. I need to use addition and subtraction to solve this problem.

2 Write the first part of the calculation you have to do.

$5,632 + 329 =$

3 Do the calculation.

$$\begin{array}{r} 5\ 6\ 3\ 2 \\ +\ \ \ 3\ 2\ 9 \\ \hline 5\ 9\ 6\ 1 \\ {\scriptstyle 1} \end{array}$$

4 Write the next part of the calculation.

$9,802 - 5,961$

5 Do the calculation.

6 What is your answer?

My answer is 3,841.

7 Check your answer by putting it into the missing number problem.

$5,632 + 3,841 + 329 = 9,802$

Try this

1 $\boxed{} + 532 + 5,035 + 1,788 = 8,073$

Top tip

- Check your answer by doing the inverse operation.

Squares and cubes

To achieve the higher score, you need to:

★ recognise and use **cube numbers**, and use the notation for cubed (^{3})

★ solve problems involving multiplication and division using your knowledge of factors and multiples, squares and cubes.

What you need to know

- A **square number** is made by multiplying a number by itself. 16 is a square number: $4 \times 4 = 16$. 16 can be written as 4^2.

- A **cube number** is made by multiplying a number by itself twice. 64 is a cube number: $4 \times 4 \times 4 = 64$. 64 can be written as 4^3.

Let's practise

Amy chooses a number between 15 and 50
It is 5 more than a square number and also 3 more than a cube number. What is her number? ☐

1 Read the question and read it again. What is it asking?

I need to list square and cube numbers between 15 and 50.
Then I need to add 5 to the square numbers and 3 to the cube numbers to find a common number.

2 List the square numbers between 15 and 50.
List the cube numbers between 15 and 50.

16; 25; 36; 49
27

3 Add 5 to the square numbers.
Also add 3 to the cube numbers.

21; 30; 41; 54
30

4 Find a number that is common to both lists.

30

5 What is your answer? Check your answer.

My answer is 30.

Top tip

- Make sure you write ^{2} to mean squared and ^{3} to mean cubed in your answers.

Try this

1 There are three square numbers greater than 200 but less than 300 What are they? ☐ ☐ ☐

2 Ravinder thinks of a number and cubes it. He gets 343 What was his number? ☐

Common multiples

To achieve the higher score, you need to:
★ identify **common multiples**.

What you need to know

- A **multiple** is a number that has been multiplied by a given number (e.g. 21 is a multiple of 7 because $7 \times 3 = 21$, and 550 is a multiple of 5 because $5 \times 110 = 550$).
- A **common multiple** is a multiple of two different numbers (e.g. $12 = 2 \times 6 = 3 \times 4$, so 12 is a common multiple of 2; 3; 4; and 6).

Let's practise

Jon chooses a common multiple of 4 and 11
It is greater than 50 but less than 150
What numbers could Jon have chosen? [   ] and [   ]

1 Read the question and read it again. What is it asking?

I need to find a common multiple of 4 and 11 that is between 50 and 150.

2 Be systematic. Work through multiples of the higher number, until you come to a multiple of the other (4).

11; 22; 33; 44
44 is a multiple of 4 too.

3 Check the number matches the criteria of the question.

The question says between 50 and 150.
44 < 50 so 44 is not an answer.

4 All multiples of 44 will also have 4 and 11 as factors. Find multiples of 44.

44; 88; 132; 176; 220 …

5 Choose a number that is > 50 and < 150. What is your answer? Check your answer.

The answer is 88 and 132.

Try this

1 Write down **three** common multiples of 4, 6 and 8 [   ] [   ] [   ]

2 Ali says, '18 is the smallest common multiple of 2 and 3 and one other number'. What is the other number? [   ]

3 Nia chooses a common multiple of 8 and 6
It is greater than 100 but less than 150
What **two** numbers could Nia choose? [   ] and [   ]

Top tip

- Questions that ask what the number *could be* mean there is more than one answer.

Common factors

To achieve the higher score, you need to:
* ★ find all the **factor** pairs of a number
* ★ solve problems involving multiplication and division using your knowledge of factors.

What you need to know

* A **factor** is a number that divides into a given number without leaving a remainder.
* A **common factor** is a factor for two different numbers
  (e.g. 2 is a common factor of 6 and 14).

Let's practise

144 This three-digit number has 8 and 9 as factors. Find another three-digit number that has 8 and 9 as factors. ☐

1 Read the question and read it again. What is it asking?

Find a three-digit number, other than 144, that has 8 and 9 as factors.

2 Find the lowest common multiple of 8 and 9.

Multiples of 9:
9; 18; 27; 36; 45; 54; 63; 72

72 is the first multiple of 9 that is also a multiple of 8

3 All multiples of 72 must also be multiples of 8 and 9. Choose any three-digit answer except for 144.

Multiples of 72:
72; 144; 216; 288; 360 ...

4 What is your answer?

Any of 216; 288; 360 ...

5 Check your answer can be divided by 8 and 9.

Top tip

* Work systematically when trying to find common factors.

Try this

1 3 and 8 are two factors of 144
   a) Find another three-digit number that has 3 and 8 as factors. ☐
   b) Find a four-digit number that has 3 and 8 as factors. ☐

2 Max thinks of a two-digit number. Max says, 'My number only has even numbers as factors.' Explain why Max is **incorrect**.

_____

Prime numbers and prime factors

To achieve the higher score, you need to:
* ★ identify and use **prime numbers** up to 100
* ★ recognise and use **prime factors**.

 What you need to know

* A **prime factor** is a factor that is a **prime number**.
* The prime factors of a number are the prime numbers that multiply to give that number.

Let's practise

> Write 24 as a product of its prime factors.
> _____

1 Read the question and read it again. What is it asking?

List the prime numbers that multiply to give 24.

2 Use a tree diagram to help you find the answer.
Find a pair of factors of 24, e.g. 2 and 12.
Only 2 is prime, so continue with a pair of factors for 12.
Again 2 is prime, so continue with a pair of factors for 6.

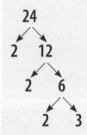

24
2   12
2   6
2   3

3 What is your answer?

The prime factors of 24 are 2, 2, 2 and 3.

4 Check the prime factors multiply to give 24.

$2 \times 2 \times 2 \times 3 = 24$
This can be written as $2^3 \times 3$.

Try this

1 Write the following numbers as a product of their prime factors.

a) 18 _____

b) 28 _____

2 Write the next four prime numbers after 75

 Top tip

* To find prime factors, start by using the smallest possible prime number.

Multiplying and dividing by 10, 100 and 1,000

To achieve the higher score, you need to:

★ multiply and divide whole numbers and those involving decimals by 10; 100; and 1,000, giving answers up to three decimal places.

What you need to know

- Dividing by 10 makes each digit worth 10 times less (e.g. 300 ÷ 10 = 30).
- Dividing by 100 makes each digit worth 100 times less (e.g. 300 ÷ 100 = 3).
- Dividing by 1,000 makes each digit worth 1,000 times less (e.g. 300 ÷ 1,000 = 0.3).

Let's practise

$$638 \div 1{,}000 = \boxed{}$$

1	Read the question and read it again. What is it asking?	**Divide 638 by 1,000.**
2	What do you know about dividing by 10?	**Dividing by 10 makes each digit worth 10 times less. 638 ÷ 10 = 63.8**
3	Dividing by 1,000 is the same as dividing by 10 three times.	**638 ÷ 10 = 63.8** **63.8 ÷ 10 = 6.38** **6.38 ÷ 10 = 0.638**
4	What is your answer?	**My answer is 0.638.**
5	Check your answer for size. Does it sound about right?	**638 is less than 1,000.** **Dividing 638 by 1,000 will give a number less than 1.**

Try this

1 a) 9,207 ÷ 10 = ☐   b) 568 ÷ ☐ = 0.568

   c) 0.306 × 1,000 = ☐   d) 42.8 × ☐ = 4,280

2 Zoe multiplies 0.103 by 1,000
   She says, 'My answer is 130'.
   Explain what Zoe has done **wrong**.

Top tip

- You can draw a **place value** chart to check if your answer is correct.

Long multiplication

To achieve the higher score, you need to:
★ solve complex problems using the written method of **long multiplication**.

⭐ What you need to know

- **Long multiplication** is done in two parts: first by multiplying the ones, and then by multiplying the tens. Add the numbers to find the answer.

⭐ Let's practise

A factory has 1,478 boxes of beans.
There are 32 tins of beans in each box.
How many tins of beans is this altogether? ☐

1 Read the question and read it again.
What is it asking?

Multiply 1,478 by 32.

2 Estimate the answer.

1,478 is about 1,500. 32 is about 30.
$1,500 \times 30 = 45,000$
The answer will be about 45,000.

3 The multiplication is done in two parts. First multiply by the ones.

```
    1 4 7 8
  ×     3 2
    2 9 5 6
      1 1
```
$1,478 \times 2 = 2,956$

4 Next, multiply by the tens. Place a zero in the ones column to represent multiplying by 10, and then multiply by the 3.

```
    1 4 7 8
  ×     3 2
    2 9 5 6
  4 4 3 4 0
    1 2 2
```
$1,478 \times 3 = 4,434$, so
$1,478 \times 30 = 44,340$

5 Add the two together.
What is your answer?

```
      1 4 7 8
    ×     3 2
      2 9 5 6
  + 4 4 3 4 0
    4 7 2 9 6
          1
```

6 Check your answer is close to your estimate.

47,296 is close to 45,000.

⭐ Try this

1 It is 1,161 km from London to Rome. A plane flies there and back four times a month. How far does the plane fly in a year?

☐

⭐ Top tip

- Don't forget the placeholder zero when multiplying by the tens digit.

Long division

To achieve the higher score, you need to:

★ use **long division** to solve problems and interpret remainders in context

★ use written division methods where the answer has up to two decimal places.

What you need to know

- The short division method can also be used to divide larger numbers by two-digit numbers. Remember to show the numbers carried over.

$$\begin{array}{r} 4\ \ 9\ \ 6 \\ 12\overline{\big)\ 5\ \ 9\ \ ^{11}5\ \ ^{7}2} \end{array}$$

Let's practise

Maria buys a yearly train pass for £2,799
She uses it for a total of 45 weeks. How much does it cost her to travel by train each week? ☐

1 Read the question and read it again. What is it asking?

I have to divide 2,799 by 45.

2 Look at the numbers and make an estimate.

2,800 ÷ 50 is a bit more than 56.

3 Start by dividing 279 by 45 to find how many times 45 goes into 279.
Write 6 in the answer space and record 270 (45 × 6) below 279 so the remainder can be found.

$$\begin{array}{r} 6 \\ 45\overline{\big)\ 2\ \ 7\ \ 9\ \ 9} \\ 2\ \ 7\ \ 0 \\ \hline 9 \end{array}$$

4 Bring down the digit 9.
Divide 99 by 45 to find out how many times 45 goes into 99.
Write 2 in the answer space and record 90 (45 × 2) below 99 so the remainder can be found.

$$\begin{array}{r} 6\ \ 2 \\ 45\overline{\big)\ 2\ \ 7\ \ 9\ \ 9} \\ 2\ \ 7\ \ 0\ \downarrow \\ \hline 9\ \ 9 \\ 9\ \ 0 \\ \hline 9 \end{array}$$

5 A remainder must be shown as a decimal for the context of money.
As no more whole 'lots of 45' go into 9, put a decimal point and a zero next to 2,799 and in the answer space.
Bring down the zero as before.

$$\begin{array}{r} 6\ \ 2\ .\ 2 \\ 45\overline{\big)\ 2\ \ 7\ \ 9\ \ 9\ .\ 0} \\ 2\ \ 7\ \ 0\ \downarrow \\ \hline 9\ \ 9 \\ 9\ \ 0\ \ \downarrow \\ \hline 9\ \ 0 \end{array}$$

6 What is your answer? Use two decimal places for money.

£62.20

Try this

1 2,822 T-shirts are packed into boxes of 28: how many **full** boxes will there be? ☐

Correspondence

To achieve the higher score, you need to:

★ solve **correspondence** problems in which *n* objects are connected to *m* objects.

What you need to know

- **Correspondence** problems can be solved using multiplication and division.

Let's practise

Ben goes on holiday. He takes four T-shirts, three pairs of shorts and two hats.

How many different combinations are there for Ben to wear his clothes?

1 Read the question and read it again. What is it asking?

I have to work out the number of ways Ben can combine his clothes.

2 It might help to make jottings. Think about the number of different pairs of shorts Ben could wear with just one T-shirt.

T-shirt 1: Ben could wear each of the 3 pairs of shorts, so that gives Ben 3 different combinations.

3 Now, think about the other T-shirts.

Ben could wear each of the 3 pairs of shorts with T-shirts 2, 3 and 4.
This gives a total of 12 combinations (4 T-shirts × 3 shorts).

4 Now, think about the hats.

Hat 1: There are 12 different combinations of T-shirts and shorts that Ben can wear with hat 1.
Two hats extends the combinations to 24 (i.e. 4 × 3 × 2).

5 What is your answer?

24 combinations

6 Check your answer.

Top tip

- Always double check your calculations.

Try this

1 Dhruv has a choice of mobile phone contracts. There is a choice of 6 phones, with 3 different allowances for minutes, 4 different allowances for texts and 3 different allowances for Internet usage. How many different choices does Dhruv have?

Order of operations

To achieve the higher score, you need to:
★ know the **order of operations** in which to carry out calculations involving the four operations and brackets.

What you need to know

- You must complete the calculations in a special order (BIDMAS). Remember × and ÷ can be done in any order and + and − can be done in any order. We must work from left to right.

- Use **BIDMAS** to remember the **order of operations**:

B	brackets
I	indices (5^2, the ^{2} is an indice)
D	division
M	multiplication
A	addition
S	subtraction

Let's practise

$$45 + 32 \times (12 - 8) \div 2 = \boxed{}$$

1   Read the question and read it again. What is it asking?

Complete the calculation using BIDMAS.

2   Remember BIDMAS. Complete any calculation in brackets.

$$45 + 32 \times (12 - 8) \div 2$$
$$= 45 + 32 \times 4 \div 2$$

3   Complete any division or multiplication. Remember to work from left to right.

$$= 45 + 128 \div 2$$
$$= 45 + 64$$

4   Complete any addition and subtraction. Again, work from left to right.

$$= 109$$

5   What is your answer? Check your answer.

109

Try this

1 $100 \div (56 - 36) + 25 \times 5 \div 10 = \boxed{}$

2   Insert a pair of brackets to make this true.

$$10 \times 10 \times 10 - 10 = 900$$

Top tips

- BODMAS can also be used. The letters stand for Brackets, Order, Division, Multiplication, Addition, Subtraction.
- If a calculation only uses addition and subtraction **or** only uses multiplication and division, work from left to right.

Ordering fractions

To achieve the higher score, you need to:

★ compare and order **fractions**, including fractions greater than 1.

What you need to know

- A fraction is used to express *part of a whole*. **Unit fractions** (e.g. $\frac{1}{2}$, $\frac{1}{3}$ and $\frac{1}{4}$) all have a **numerator** of 1. **Proper fractions** (e.g. $\frac{3}{4}$ and $\frac{5}{6}$) are less than or equal to 1 whole.
- Fractions greater than 1 are called **improper fractions** and can also be written as a **mixed number** (e.g. $\frac{9}{8} = 1\frac{1}{8}$).

Let's practise

Write these fractions in order, starting with the **smallest**.

$$2\frac{1}{6} \qquad \frac{5}{8} \qquad 2\frac{2}{3}$$

smallest ☐ ☐ ☐ largest

1 Read the question and read it again. What is it asking?

Write the fractions in order. I have to start with the smallest.

2 First check if any fractions do not have whole numbers. These will be the smallest.

Only $\frac{5}{8}$ doesn't have any whole numbers and it is less than 1, so it must be the smallest.

3 Now check the numbers that have whole numbers. Use common multiples to compare fractions with different denominators.

Two mixed numbers have 2 as a whole number, so I need to compare their two fractions.
$\frac{2}{3} = \frac{4}{6}$, so $2\frac{1}{6}$ must be less than $2\frac{2}{3}$.

4 Now the fractions can be put in order.

$$\frac{5}{8} \qquad 2\frac{1}{6} \qquad 2\frac{2}{3}$$

5 Check your answer. Make sure you started with the smallest.

Top tip

- Use **equivalent fractions** to make the denominators equal.

Try this

1 Put these fractions in order, starting with the **largest**.

a) $\frac{2}{3}$ $\frac{3}{5}$ $\frac{7}{10}$ largest ☐ ☐ ☐ smallest

b) $4\frac{2}{3}$ $4\frac{7}{12}$ $3\frac{5}{6}$ $4\frac{5}{8}$ largest ☐ ☐ ☐ ☐ smallest

Adding and subtracting fractions

To achieve the higher score, you need to:
★ add and subtract fractions with different **denominators** and mixed numbers, using the concept of **equivalent fractions**.

What you need to know

- Denominators in fractions need to be equal before they can be added or subtracted. Use equivalent fractions that have the same value even though they look different (e.g. $\frac{1}{2} = \frac{2}{4}$).
- Simplifying fractions involves dividing the numerator and denominator by a common factor.

Let's practise

$$6\frac{1}{6} - 2\frac{5}{8} = \boxed{}$$

1 Read the question and read it again. What is it asking?

Subtract two and five-eighths from six and one-sixth.

2 Find the common denominator.
To complete $\frac{4}{24} - \frac{15}{24}$, take one whole from the 6, leaving 5.
$$\frac{24}{24} + \frac{4}{24} - \frac{15}{24} = \frac{13}{24}$$

The common denominator is 24.
$$6\frac{1}{6} = 6\frac{4}{24} \qquad 2\frac{5}{8} = 2\frac{15}{24}$$
Now, they can be subtracted.
$$6\frac{4}{24} - 2\frac{15}{24} = 3\frac{13}{24}$$

3 What is your answer?

$$3\frac{13}{24}$$

4 Check your answer using addition.

$$2\frac{5}{8} + 3\frac{13}{24} = 6\frac{1}{6}$$

Try this

1 $3\frac{2}{3} + 4\frac{3}{8} = \boxed{}$

2 $3\frac{1}{4} - 1\frac{4}{5} = \boxed{}$

3 $4\frac{2}{5} - 2\frac{5}{6} = \boxed{}$

4 $5\frac{7}{8} + 3\frac{1}{6} = \boxed{}$

Top tip

- Always check your answer to a subtraction calculation by adding what was taken away to your answer.

21

Multiplying fractions

To achieve the higher score, you need to:

★ multiply **proper fractions** and mixed numbers by whole numbers

★ multiply simple pairs of proper fractions, writing the answer in its simplest form.

What you need to know

- When you multiply **fractions** you are working out a fraction of a fraction.

Let's practise

$$\frac{4}{5} \times \frac{1}{2} = \boxed{}$$

1 Read the question and read it again. What is it asking?

I have to multiply four-fifths by one-half.

2 First, imagine a diagram showing $\frac{4}{5}$.

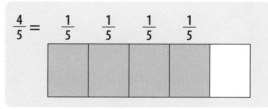

3 $\frac{4}{5} \times \frac{1}{2}$ is the same as finding $\frac{1}{2}$ of $\frac{4}{5}$.

We can show this on the diagram by splitting each fifth into two halves and shading one of these halves. This area is shown in a darker shade.

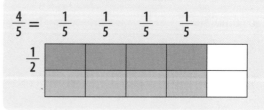

4 We can now see the darker shade represents $\frac{4}{10}$

so $\frac{4}{5} \times \frac{1}{2} = \frac{4}{10}$ or $\frac{2}{5}$.

$\frac{4}{10} = \frac{2}{5}$

Top tip

- To multiply fractions by whole numbers, turn the whole number into a fraction, e.g. $\frac{2}{3} \times 4 = \frac{2}{3} \times \frac{4}{1}$.

5 When you understand what is happening, you can use a quicker method. Multiply the numerators together and multiply the denominators together.

$\frac{4}{10} = \frac{2}{5}$

Try this

1 What is three-eighths of three-quarters? $\boxed{}$

2 What is three-quarters of two-sevenths? $\boxed{}$

3 Calculate:

a) $\frac{3}{8} \times 5 = \boxed{}$

b) $2\frac{1}{4} \times 4 = \boxed{}$

Dividing fractions

To achieve the higher score, you need to:
★ divide **proper fractions** by a whole number.

What you need to know

- A whole number is a number without any fractions (e.g. 3; 17; 206). A proper fraction is a fraction where the numerator is less than the denominator (e.g. $\frac{2}{3}$, $\frac{5}{12}$, $\frac{9}{10}$).
- Division facts are useful when dividing fractions.

Let's practise

$$\frac{1}{4} \div 5 = \boxed{}$$

1 Read the question and read it again. What is it asking?

I have to divide one-quarter by 5.

2 First, imagine a diagram showing $\frac{1}{4}$.

3 The quarter is divided into five parts.

4 If each quarter were divided in the same way, there would be 20 parts. So each part must be $\frac{1}{20}$.

5 What is your answer?

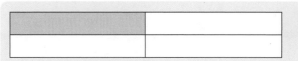

$$\frac{1}{4} \div 5 = \frac{1}{20}$$

6 Check your answer using an inverse calculation.

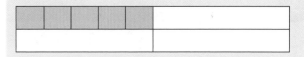

$$\frac{1}{20} \times 5 = \frac{1}{4}$$

Try this

1 a) $\frac{1}{3} \div 4 = \boxed{}$ b) $\frac{1}{4} \div 8 = \boxed{}$

2 Find the missing number.

a) $\frac{1}{5} \div \boxed{} = \frac{1}{25}$ b) $\frac{1}{8} \div \boxed{} = \frac{1}{24}$

Top tip

- Look for the multiplication pattern in these calculations.

Changing fractions to decimals

To achieve the higher score, you need to:
★ use division to calculate decimal fraction equivalents.

What you need to know

- In decimal numbers, the decimal point separates the whole number part from the fractional part.

- The first three digits after the decimal point represent tenths, hundredths and thousandths.

Let's practise

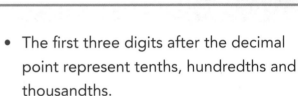

Write $1\frac{3}{8}$ as a decimal. ☐

1 Read the question and read it again. What is it asking?

$1\frac{3}{8}$ is a mixed number. I have to write it as a decimal.

2 What do you know about mixed numbers?

$1\frac{3}{8} = 1 + \frac{3}{8}$

3 What does $\frac{3}{8}$ mean?

$\frac{3}{8}$ means 3 divided into 8 parts.

4 Do the division: $3 \div 8$. Remember to put in the decimal point and to add zeros when you have a remainder.

$$8 \overline{\smash{)}3 . {}^3 0 \, {}^6 0 \, {}^4 0} = 0 . 3 \, 7 \, 5$$

5 What is your answer? Don't forget the whole number part.

My answer is 1.375.

6 Check your answer using the inverse calculation.

$1.375 = \frac{1,375}{1,000} = 1\frac{3}{8}$

Try this

1 Write $3\frac{1}{2}$ as a decimal. ☐

2 Write $\frac{1}{8}$ as a decimal. ☐

3 Write $4\frac{5}{8}$ as a decimal. ☐

Top tip

- To turn a fraction into a decimal, divide the numerator by the denominator.

Rounding decimals

To achieve the higher score, you need to:

★ round decimals with up to three decimal places to the nearest whole number, or nearest one, or two, decimal places.

What you need to know

- Decimal fractions can be rounded to the nearest whole number or to a number of **decimal places**.
- A decimal place is the position of a digit to the right of the decimal point.
- If a digit is 5 or above, you have to round up. If a digit is 4 or less, you have to round down.

Let's practise

Round 13.638 to one decimal place.

| ① | Read the question and read it again. What is it asking? | Round 13.638 to one decimal place. |

| ② | Make sure you know which digits are in the different decimal places. | 1 3 . 6 3 8
        1st 2nd 3rd decimal places |

| ③ | Round 13.638 to one decimal place. 13.638 comes between 13.6 and 13.7. Check the second decimal place. | 1 3 . 6 3 8
        ↑
 The second decimal place is a 3, so I have to round down. |

| ④ | What is the answer? | 13.6 |

| ⑤ | Check you have rounded to the number of decimal places asked for. | |

Top tip

- Make sure you know which digits are in the first, second and third decimal places.

Try this

1 Round 25.729 to: a) one decimal place. ☐   b) two decimal places. ☐

2 Tick (✓) any numbers that could be rounded to 37.6

☐ 38.002   ☐ 37.638   ☐ 37.58   ☐ 37.71   ☐ 37.661

Adding and subtracting decimals

To achieve the higher score, you need to:
★ add and subtract decimals with up to three decimal places, and numbers with different numbers of decimal places.

What you need to know

- Line up the decimal points when adding or subtracting decimal numbers.

Let's practise

$$78.537 + 6.85 = \boxed{}$$

1	Read the question and read it again. What is it asking?	I have to add the two decimal numbers.
2	Estimate the answer.	Rounding to the nearest 10: 78.537 is near 80 and 6.85 is near 10. $80 + 10 = 90$, so the answer will be around 90.
3	Set out the calculation, lining up the decimal points. Gaps beyond the decimal point should be filled with zeros.	$\begin{array}{r} 7\,8\,.\,5\,3\,7 \\ +\quad 6\,.\,8\,5\,0 \\ \hline \end{array}$
4	Do the sum.	$\begin{array}{r} 7\,8\,.\,5\,3\,7 \\ +\quad 6\,.\,8\,5\,0 \\ \hline 8\,5\,.\,3\,8\,7 \\ {\scriptstyle 1\ 1} \end{array}$
5	What is your answer? Remember to include the decimal point in your answer.	85.387
6	Check your answer against your estimate.	My answer is close to 90.

Try this

1 74.581 + 26.855 = ☐

2 45.653 – 7.808 = ☐

3 68.43 + 7.837 = ☐

4 84.165 – 76.42 = ☐

Top tip

- Gaps beyond the decimal point can be filled with zeros. This is especially useful when doing a subtraction.

Multiplying decimals

To achieve the higher score, you need to:

★ multiply one-digit numbers with up to two decimal places by one-digit and two-digit whole numbers.

What you need to know

- Multiplication facts are useful when multiplying decimals.
- It is useful to know if a number has been multiplied or divided by 10 or 100 when multiplying decimals.

Let's practise

$$0.07 \times 28 = \boxed{}$$

1 Read the question and read it again. What is it asking?

I have to multiply 0.07 by 28.

2 Partition the calculation.

0.07×20
0.07×8

3 Use the key facts of 7×8 and 7×2. But remember that 0.07 isn't 7, it's $7 \div 100$. And 20 isn't 2, it's 2×10.

I can think of the calculations as:
$7 \div 100 \times 8$ or $7 \times 8 \div 100$
$7 \div 100 \times 2 \times 10$ or $7 \times 2 \div 100 \times 10$

4 Add the two numbers.

$$0.07 \times 8 = 0.56$$
$$0.07 \times 20 = \underline{1.4}$$
$$1.96$$

5 Write your answer. Don't forget the decimal point.

1.96

6 Check your answer using a calculation.

$1.96 \div 0.07 = 1.96 \div 7 \times 100 = 28$

Try this

1 $0.05 \times 7 = \boxed{}$

2 $0.6 \times 8 = \boxed{}$

3 $0.09 \times 4 = \boxed{}$

4 $0.9 \times 9 = \boxed{}$

5 $0.3 \times \boxed{} = 2.1$

6 $\boxed{} \times 5 = 0.45$

7 $\boxed{} \times 0.06 = 0.36$

8 $\boxed{} \times 0.7 = 2.8$

Top tip

- Remember to use your knowledge of dividing numbers by 10 and 100.

Percentages

To achieve the higher score, you need to:

★ solve problems involving the calculation of **percentages** and use percentages for comparison

★ compare percentages and fractions including tenths, fifths, halves and quarters.

What you need to know

- **Percentages** are used to describe **proportion**.
- The percentage symbol (%) represents a fraction with a denominator of 100. So, 25% means $\frac{25}{100}$.

Let's practise

There are 200 people at a concert. A quarter of them are children and 62% are women.

How many men are at the concert? ☐

1 Read the question and read it again. What is it asking?

I need to work out how many men are at the concert by subtracting the number of children and women from the total.

2 Find a quarter of 200 to find the number of children.

$200 \div 4 = 50$ (children)

3 Remember what % means and find the fraction equivalent. Divide by 100 to find 1%. Multiply by 62 to find 62%.

% means out of 100 so
$62\% = 62$ out of 100 or $\frac{62}{100}$
$200 \div 100 = 2$
$2 \times 62 = 124$ (women)

4 Subtract the number of children and women from the total number of people at the concert.

$50 + 124 = 174$
$200 - 174 = 26$

5 What is your answer?

There are 26 men at the concert.

6 Check your answer.

$50 + 124 + 26 = 200$

Try this

1 Callum saves 15% of his pocket money. He gets £2 each week. How much does he save in one year? ☐

2 Millie buys a coat in a sale. The sale price is £64. The coat is 20% off. What was the starting price of the coat? ☐

Ratio

To achieve the higher score, you need to:

★ solve **ratio** problems where missing values can be found by using multiplication and division facts.

What you need to know

- **Ratio** compares the relative sizes of quantities or numbers.

Let's practise

> At school, children can have a packed lunch or a school dinner. The ratio of children who have a packed lunch to those who have a school dinner is **2:7**
>
> 168 children have a school dinner.
>
> How many children eat at school altogether? ☐

1 Read the question and read it again. What is it asking?

Work out how many children have a packed lunch and add it to the number of children who have a school dinner.

2 Think of ratio as shares. School dinners are 7 shares because the ratio is 2 (packed lunch) : 7 (school dinner)

168 children have a school dinner. $168 \div 7 = 24$ Each share is worth 24.

3 Packed lunches are 2 shares.

$24 \times 2 = 48$

4 Add the two numbers together.

$168 + 48 = 216$

5 What is your answer?

216 children eat at school altogether.

6 Check your answer.

Try this

1 The ratio of boys to girls at a school is **5:4**
There are 165 boys.
How many children are in the school? ☐ children

2 The ratio of dogs to cats in an animal shelter is **4:3**
There are 23 more dogs than cats.
How many cats are there? ☐ cats

Top tip

- **7:2** is not the same as **2:7**. Read the question twice to make sure you are using the correct numbers in your calculation and providing the answer in the order required.

Proportion

To achieve the higher score, you need to:

★ solve problems involving the relative sizes of two quantities.

What you need to know

- Proportion compares different fractions or ratios.

Let's practise

Two eggs are used for every 400 g of flour to make Yorkshire puddings.

How many eggs are needed if 2,000 g of flour is used?

| | eggs |

1 Read the question and read it again. What is it asking?

Find out how many eggs are needed for 2,000 g of flour.

2 Work out the ratio of eggs to grams of flour. Simplify it if possible.

2:400
1:200 (simplified)

3 So, every 200 g of flour needs 1 egg. What is 2,000 divided by 200?

2,000 ÷ 200 = 10

4 What is your answer?

10 eggs

5 Check your answer using a different strategy.

If 400 g uses 2 eggs, 800 g uses 4 eggs ... 2,000 g uses 10 eggs.

Try this

1 A jar of 200 nails has a **mass** of 2.5 kg.
What is the mass of 1,000 nails? | | kg

2 Four out of every nine passengers on a train are female.

a) There are 120 female passengers.
How many passengers are there altogether? | | passengers

b) What **fraction** of the passengers are male? | |

c) What is the **ratio** of female passengers to male passengers? | |

Scaling problems

To achieve the higher score, you need to:
★ solve scaling problems, including scaling by simple fractions
★ solve problems where the **scale factor** is known or can be found.

What you need to know

- Scaling problems are about keeping numbers and amounts in proportion.

Let's practise

Triangle B is an enlargement of triangle A.
How long is the side X?

7.5 cm
A
Not to scale
12 cm

X
B
48 cm

1 Read the question and read it again. What is it asking?	Work out how long side X is in centimetres.
2 Triangle B is an enlargement of triangle A. Look at what you know first.	If triangle B is an enlargement of triangle A, then the triangles are the same shape but a different size. The base side has increased from 12 cm to 48 cm.
3 Think how many times the side has increased.	$48 \div 12 = 4$ So, the base side is 4 times bigger. This is called the scale factor.
4 Use the scale factor to enlarge 7.5 cm.	$7.5 \times 4 = 30$
5 What is your answer?	$X = 30$ cm
6 Check your answer.	

Top tip

- Scale factors can be fractions as well as numbers greater than 1.

Try this

1 24 tins have a total mass of 5,400 grams. What is the mass of 60 tins? ☐ kg

2 A rectangle is 18 cm long and 5 cm wide. It is enlarged by a scale factor of **1.5**

What will the **new** length and width be? Length = ☐ cm   Width = ☐ cm

Unequal sharing

To achieve the higher score, you need to:

★ solve problems involving **unequal sharing** and grouping using your knowledge of fractions and multiples.

What you need to know

- **Sharing equally** involves dividing the total amount by the number of groups (e.g. £25 shared into two groups is £12.50 each).
- With **unequal sharing**, the total number or quantity is divided into different proportions, fractions or **percentages**.

Let's practise

£100 is to be shared by three brothers in the ratio 2:1:1
How much does each get?

1 Read the question and read it again. What is it asking?

Divide £100 between three brothers in the ratio given.

2 What does 2:1:1 mean?

The first brother is to receive twice as much as each of the other two brothers.

3 Add the numbers to find out how many equal shares there are. Then work out the value of each equal share.

$2 + 1 + 1 = 4$
£100 ÷ 4 = £25
Each share is worth £25.

4 Distribute the shares according to the ratio given (to create the unequal sharing).

$2 \times 25 = 50, 1 \times 25 = 25$
So they get £50, £25 and £25, respectively.

5 Check your answer by adding the shares back together.

£50 + £25 + £25 = £100

Try this

1 Share £1,000 so that Jai gets 63%, Hope gets 24% and Mia gets 13%.

Jai [ ] Hope [ ] Mia [ ]

2 Eddie and Oscar share a pizza. Eddie eats four times as much as Oscar.

What **fraction** does Oscar eat? [ ]

Algebra

To achieve the higher score, you need to:
★ express missing number problems using **algebra**.

What you need to know

- In **algebra**, words or **symbols** are used to represent the amounts in a problem, instead of actual numbers.

- An **equation** with letters can help to solve a missing number problem (e.g. $x + 5 = 9$, so $x = 4$).

Let's practise

Jessica thinks of a number. She multiplies it by 4 and adds 7; her answer is 31

Write an algebraic equation that shows this.

1	Read the question and read it again. What is it asking?	Write an equation for Jessica's missing number calculation.
2	Use *y* for the number Jessica thought of. (Any letter can be used.)	$y \times 4 + 7 = 31$
3	In algebra, write $y \times 4$ as $4y$.	$4y + 7 = 31$
4	Check the equation you have written matches the question.	**a number is multiplied by 4** $4 \times y = 4y$ **7 is added** $4y + 7$ **the answer is 31** $4y + 7 = 31$

Try this

For each of these problems, write an equation and solve it.

1 Georgie thinks of a number.

She multiplies it by 2, then subtracts 7 and gets 51

a) Write an algebraic equation that shows this.

b) What was her number?

2 Lottie thinks of a number.

She divides it by 2, then multiplies by 3 and gets 27

a) Write an algebraic equation that shows this.

b) What was her number?

Top tip

- To solve algebra problems, work backwards and use inverse operations.

Using formulae

To achieve the higher score, you need to:

★ use and write simple **formulae** using algebra.

What you need to know

- A **formula** has unknown numbers in it. These are usually written as letters.
- Once you know the value of one of the letters, you can substitute the actual number for that letter.

Let's practise

The formula for finding the height (h) of a triangle when you know the area (A) and the length of the base (b) is:

$$h = \frac{2A}{b}$$

Use the formula to find the height of a triangle that has an area of 39 cm² and a base of 12 cm. ☐ cm

1 Read the question and read it again. What is it asking?

Work out the height of the triangle using the formula given.

2 Use the formula and the values you know:
$A = 39$
$b = 12$

I will need to substitute the values I know into the formula.

$h = \frac{2A}{b}$ $\qquad$ $h = \frac{2 \times 39}{12}$

3 Now calculate.

$h = \frac{78}{12} = 6.5$

4 What is the answer?

The height of the triangle is 6.5 cm.

5 Check your answer by rearranging the equation.

Top tip

- When replacing an unknown letter with a number, make sure the units of the number match the units used in the formula.

Try this

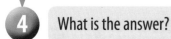

1 A plumber works out her fees for a job using this formula: $F = 25h + 0.2k$

F is the fee, h is the number of hours she works and k is the number of kilometres she must drive to get to the job. Work out the plumber's fee for a job that takes 5 hours and is 40 kilometres from her home. ☐

Solving equations

To achieve the higher score, you need to:
* ★ find a pair of numbers that satisfies an **equation** with two unknowns
* ★ calculate possible combinations of two **variables**.

What you need to know

* The = sign in an equation shows that both sides must balance.

Let's practise

$\bigcirc + \square = 15$ $\bigcirc \times \square = 36$

These shapes stand for two different whole numbers.
Find the value of each shape.

$\bigcirc = \boxed{}$ $\square = \boxed{}$

1 Read the question and read it again. What is it asking?

Each shape stands for a number. I need to find two values that make both calculations correct.

2 Begin with the first calculation. List what the values could possibly be.
At the same time, check to see if they also multiply to 36.

I know that when the two numbers are added they total 15 and when the same two numbers are multiplied they come to 36.
$15 + 0 = 15, 15 \times 0 = 0$
$14 + 1 = 15, 14 \times 1 = 14$
$13 + 2 = 15, 13 \times 2 = 26$
$12 + 3 = 15, 12 \times 3 = 36$

3 When you've found a pair that give the right answers, the problem is solved.

$12 + 3 = 15$ and $12 \times 3 = 36$

4 What is the answer?

$\bigcirc = 12$ $\square = 3$

(In this question, the circle could be 12 or 3 and the square could be 12 or 3.)

5 Check your answer by putting the values back in the calculations.

Try this

1 Find the value of the shapes.

$\triangle \times \diamond = 40$ and $\triangle - \diamond = 3$

$\triangle = \boxed{}$ $\diamond = \boxed{}$

2 Find the value of the letters.

$R \div T = 5$ and $R - T = 16$

$R = \boxed{}$ $T = \boxed{}$

Linear number sequences

To achieve the higher score, you need to:

★ generate, describe and complete **linear number sequences**.

What you need to know

- In a **linear number sequence**, numbers follow a pattern according to a **rule** (or a formula). The rule can be described by an expression using **n**.
- Finding the **nth term rule** lets you find any number in the sequence.

Let's practise

Gita makes four patterns using squares.

Pattern 1   Pattern 2   Pattern 3   Pattern 4

Write a rule to work out the number of squares in any pattern. _____

1 Read the question and read it again. What is it asking?

I need to write a rule to describe how the pattern grows.

2 Count the squares to find the numbers being used.

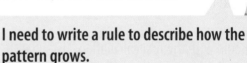

Pattern	1	2	3	4
Squares	3	5	7	9

The sequence goes up in twos.

3 Use n instead of the number of the pattern. Relate the pattern number n to the number of squares. Look at pattern 1.

If n is the pattern number (1) and the squares increase by multiples of 2, I must start with $2n$ ($2 \times 1 = 2$). But pattern 1 has 3 squares so I have to add 1 to get the correct number of squares.

4 What is the answer?

$2n + 1$

5 Check your rule works with the patterns shown.

Pattern 3: $(2 \times 3) + 1 = 7$
Pattern 4: $(2 \times 4) + 1 = 9$
My rule works.

Try this

1 Find the rule as an expression using n to describe this sequence: 6 11 16 21 26 _____

2 Write the first **five** terms of a sequence that uses the rule $3n + 5$ ⬜⬜⬜⬜⬜

Top tip

- The number each term increases gives the number to multiply n (e.g. $4n$). Then add or subtract a number to get the term (e.g. $4n - 3$).

Measures

To achieve the higher score, you need to:
* ★ solve problems involving the calculation and conversion of units of measure, using up to three decimal places.

What you need to know

* Metric conversion facts are useful when solving measures problems (e.g. 1 km = 1,000 m, 1 m = 100 cm, 1 m = 1,000 mm, 1 l = 100 cl, 1 l = 1,000 ml, 1 kg = 1,000 g).

Let's practise

Jug B is empty and jug A has some water in it.

Jug A is used to fill jug B to the top of its measure.

How much is left in jug A?

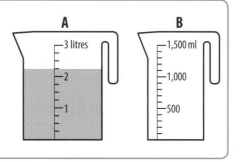

1	Read the question and read it again. What is it asking?	I have to work our how much is left in jug A after filling up jug B.
2	Be systematic. Start with jug A. How much water is in it?	The level in jug A shows 2.25 litres.
3	Now read jug B.	Jug B can hold 1,500 ml.
4	Change the units to be the same.	Jug A: 2.25 l = 2,250 ml Jug B: 1,500 ml
5	Do the calculation.	2,250 – 1,500 = 750
6	What is your answer?	My answer is 750 ml.
7	Check your answer.	

Top tip

* Always double check what each increment stands for on the scale.

Try this

Look at the jugs on this page.

1 Jug A already has 2.25 litres in it. If 625 ml of water is added to jug A, how much **more** water can jug A hold? ☐ ml

Converting metric units

To achieve the higher score, you need to:

★ convert between different units of **metric** measure

★ convert measurements of **length**, **mass**, **volume** and **time** from a smaller unit of measure to a larger unit, and vice versa, using up to three decimal places.

What you need to know

- *Kilo* stands for 1,000. *Milli* stands for $\frac{1}{1,000}$. *Centi* stands for $\frac{1}{100}$.
- 1 kilogram = 1,000 grams   1 litre = 1,000 millilitres
- 1 kilometre = 1,000 metres   1 metre = 100 centimetres   1 metre = 1,000 millimetres

Let's practise

Write 7 mm as metres. ☐ m

1	Read the question and read it again. What is it asking?	I have to convert 7 millimetres to metres.
2	What do you know about mm and m?	1 m = 100 cm, 1 cm = 10 mm
3	What is the relationship between millimetres and metres?	1 m = 100 × 10 mm = 1,000 mm 1,000 mm = 1 m
4	To change millimetres into metres, divide by 1000.	7 ÷ 1,000 = 0.007
5	What is your answer?	0.007 m
6	Check your answer by using the inverse operation.	0.007 m × 1,000 = 7 mm

Try this

1 Cream can be bought in different-sized tubs.

☐ 150 ml tub costs 45p

☐ 250 ml tub costs 60p

☐ 600 ml tub costs £1.50

Tick (✓) the tub that is the best value per 100 ml.

2 Three equal-sized pieces of cheese have a mass of 1.95 kg altogether.

The shop charges 90p per 100 g.

What is the cost of **one** piece of cheese?

☐

Metric units and imperial measures

To achieve the higher score, you need to:
* ★ understand and use approximate equivalences between metric units and common **imperial units** such as inches, pounds and pints
* ★ convert between miles and kilometres.

What you need to know

Length	1 cm ≈ 0.4 inch	1 inch ≈ 2.5 cm
	1 m ≈ $3\frac{1}{4}$ feet	1 foot (12 inches) ≈ 30 cm
Mass	1 kg ≈ 2.2 pounds	1 pound ≈ 450 g
Capacity	1 litre ≈ 1.75 pints	1 pint ≈ 570 ml

Let's practise

A recipe needs 2,000 ml of milk.
What is this as an **approximate** imperial unit? [   ]

1 Read the question and read it again. What is it asking?

I need to use the approximate conversion information and change 2,000 ml into pints.

2 Use the information you know about converting from metric to imperial units. Convert 2,000 ml into pints.

I know 1 l is about 1.75 pints. 2,000 ml is 2 litres, so I need to find 1.75 multiplied by 2.

$$\begin{array}{r} 1.75 \\ \times\ \ \ \ 2 \\ \hline 3.50 \\ {\scriptstyle 1\ \ \ 1} \end{array}$$

3 What is your answer?

2,000 ml is about 3.5 pints.

4 Check your answer.

1.75 + 1.75 = 3.50

Top tip
* Make sure you know approximate conversions for the common units, e.g. 1 inch is about 2.5 centimetres.

Try this

Use the approximate conversion information given at the top of the page.
You must give the **correct units** with each answer.

1 Mohammed needs a piece of wood that is 24 inches long. The shop only sells wood using metric measures. What length of wood should he ask for? [   ]

2 14 pounds equals 1 stone. What is the approximate metric equivalent? [   ]

3 Kim's granddad says, 'I can remember when petrol was 64p a gallon.'
What was this cost as pence per litre? (8 pints = 1 gallon) [   ]

Perimeter and area

To achieve the higher score, you need to:
★ measure and calculate the **perimeter** and **area** of composite rectilinear shapes in centimetres and metres.

What you need to know

- The **perimeter** of a shape is the total distance around the outside of a shape.
- The **area** of a shape is the total space inside a shape.
- Compound shapes may be cut into separate rectangles, then the area is worked out for each separate rectangle and finally the separate areas are added together.

Let's practise

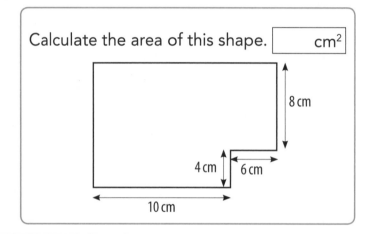

Calculate the area of this shape. ☐ cm²

8 cm

4 cm 6 cm

10 cm

1 Read the question and read it again. What is it asking?

Work out the total area of the shape.

2 Think of a method you can use.

Split the shape into two rectangles.
The rectangle A is 10 cm wide and (4 + 8) cm long.
The rectangle B is 8 cm by 6 cm.

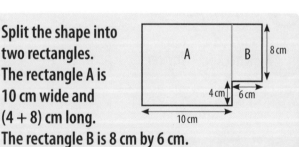

3 Use the formula $A = l \times w$ for each rectangle. Remember to use cm² to represent cm × cm.

For A: 12 cm × 10 cm = 120 cm²
For B: 8 cm × 6 cm = 48 cm²
A + B: 120 cm² + 48 cm² = 168 cm²

4 What is your answer?

The area is 168 cm².

Try this

1 Draw a rectangle with the **same area** as this compound shape.

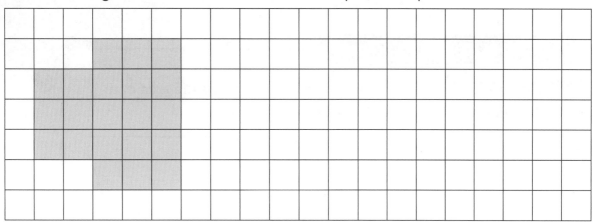

2 Tim draws two different rectangles.

Tim says, 'One rectangle has an **area** of 24 cm² and the other has a **perimeter** of 24 cm. They both have the same length.'

What could the widths of Tim's two rectangles be? ☐ cm ☐ cm

3 Work out:

a) the perimeter of this shape ☐ cm

b) the area of this shape. ☐ cm²

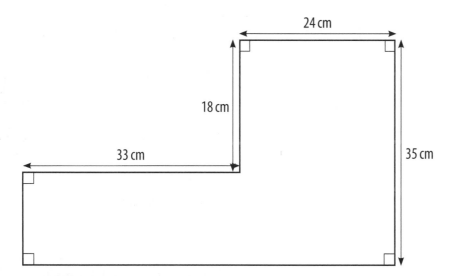

Top tips

- Make sure all the sides are in the same unit, either millimetres or centimetres.
- Remember that the units for area are always squares (e.g. cm² or m²) and the units for perimeter are always a length (e.g. cm or m).
- **Composite shapes** are the same as compound shapes.

Area of parallelograms and triangles

To achieve the higher score, you need to:

★ calculate the area of parallelograms and triangles.

What you need to know

- The formula for the area of a rectangle is the same as the area of a **parallelogram**, which is length × width (or height).

  Area of a parallelogram = length × height

  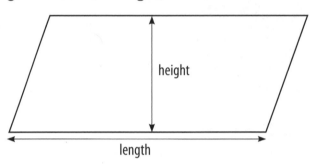

  height

  length

- The formula for the area of a **triangle** is the height × the base divided by 2

  $$A = \frac{bh}{2}$$

  Area of a triangle = base × height ÷ 2

  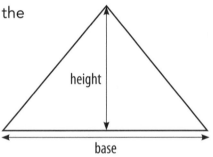

  height

  base

Let's practise

What is the area of a parallelogram that has a length of 40 cm and a height of 25 cm? ☐ cm²

1	Read the question and read it again. What is it asking?	Calculate the area of the shape.
2	Recall the formula for the area of a parallelogram.	Area of a parallelogram = length × height The parallelogram is 40 cm long and 25 cm high.
3	Calculate the area.	40 × 25 = 1,000
4	What is your answer?	1,000 cm²
5	Check your answer.	1,000 ÷ 25 = 40

Try this

Calculate the areas of these shapes.

1 Not to scale

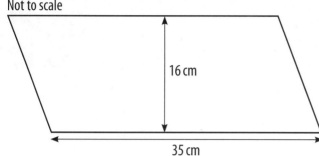

16 cm

35 cm

A parallelogram 35 cm long and 16 cm high.

☐ cm²

2 Not to scale

10 cm

20 cm

A triangle with a base 20 cm long and a height of 10 cm.

☐ cm²

3 Not to scale

15 cm

18 cm

A triangle with a base 18 cm long and a height of 15 cm.

☐ cm²

Top tips

- You don't need to write the units during your calculation, but make sure you include them in your answer if they are not given.
- The diagonal of a parallelogram creates two identical triangles. The area of each triangle is half the area of the parallelogram.

Volume

To achieve the higher score, you need to:

★ calculate, estimate and compare **volumes** of **cubes** and **cuboids** using standard units.

What you need to know

- The **volume** of a **cuboid** can be worked out using this formula:
  Volume = length × width × height.

Let's practise

Irum has some centimetre cubes.
She uses them to make a cuboid 6 cm long, 4 cm wide and 3 cm high.
How many cubes has she used? [ ]

1 Read the question and read it again. What is it asking?

I must first work out the volume of the cuboid in cubic centimetres (cm³).

2 Remember the formula. Substitute the numbers into it.

$V = l \times w \times h$
$V = 6 \times 4 \times 3$

3 Find the volume.

$6 \times 4 \times 3 = 72$
The volume is 72 cm³.

4 What is your answer?

Irum used 72 cubes.

5 Check your answer.

Try this

1 The volume of a cuboid is 450 cm³.
The area of one of the sides is 30 cm².
What could the length, width and height of the cuboid be?

Length = ___ cm   Width = ___ cm   Height = ___ cm

2 Luke has 300 cubes.

Each one is 1cm³.

He uses **all** the cubes to build a cuboid.

What could the length, width and height of Luke's cube be?

| Length =   cm | Width =   cm | Height =   cm |

3 Faith makes a cube that has a volume of 27 cm³.

She decides she wants to make a cube that has sides that are three times as long.

What will the volume of Faith's new cube be? | cm³ |

4 Irum makes a cube using centimetre cubes that has sides of 4 cm.

She uses the same number of centimetre cubes to make a cuboid.

What could the length, width and height of this cuboid be?

| Length =   cm | Width =   cm | Height =   cm |

5 Obi is using 1cm cubes to make a cube with sides of 5 cm.

He has only partly completed the cube.

How many more cubes does Obi need? | cubes |

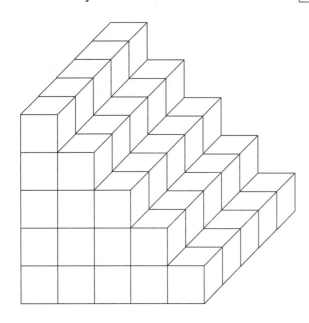

Top tip

• When giving the volume of a 3-D shape, make sure all the units are cubed (e.g. cm³ or m³).

Angles and degrees

To achieve the higher score, you need to:

★ recognise **angles** and find missing angles where they meet at a point or are on a straight line or are vertically opposite

★ find unknown angles in any triangle, quadrilateral or **regular polygon**.

What you need to know

- **Angles** that meet at a point add to 360°.
- Angles on a straight line add to 180°.
- Angles within a **triangle** add to 180°.
- Angles within a quadrilateral add to 360°.
- Angles that are vertically opposite are equal in size.

Let's practise

Calculate angle *a* and angle *b*.

a = ☐
b = ☐

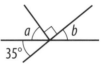

① Read the question and read it again. What is it asking?

Calculate the sizes of angles *a* and *b*.

② What do you know about the angles in the diagram? What do you know about angles on a straight line?

Angle *a* is one of three angles on a straight line. One of these angles is also a right angle (90°). They add up to 180°. They sum to 180° so 90° + 35° + *a* = 180°.
Or simply: 35° + *a* = 90°
Angle *a* = 55°

③ What do you know about angle *b*? What do you know about vertically opposite angles?

It is one of three angles on a straight line, but it is also vertically opposite the angle labelled 35°. They are equal in size so angle *b* = 35°.

④ What is your answer? Check your working.

Angle *a* = 55° Angle *b* = 35°
Angles *a* + *b* must also equal 90° as they are on a straight line with a right angle.

Try this

1 Calculate the sizes of angles *x* and *y*.

x = ☐ ° y = ☐ °

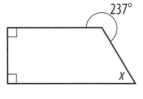

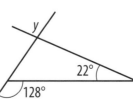

Top tip

- When lines cross, the opposite angles are equal.

Circles

To achieve the higher score, you need to:

★ illustrate and name parts of circles (including **radius**, **diameter** and **circumference**).

What you need to know

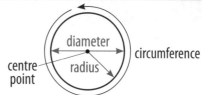

- A circle is a perfectly round 2-D shape. Every point on the **circumference** is the same distance from its **centre point**.
- The **radius** of the circle is the distance from the centre to the circumference.
- A **diameter** of the circle goes from one point on the circumference through the centre and to a point on the opposite side of the circle, cutting the circle into two equal parts.
- The radius is half the **length** of the diameter.

Let's practise

Shareen draws a pattern on centimetre-squared paper.

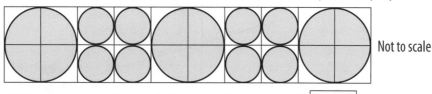

Not to scale

What is the radius of the smaller circles? ☐

1 Read the question and read it again. What is it asking?

Find the radius of the smaller circles.

2 Look at the diagram. How will you work out the radius of the smaller circles?

The small circles fit within a square of the centimetre-squared paper. So the diameter is 1 cm and the radius is half of that.

3 What is your answer? Don't forget your units.

The radius of the small circle is 0.5 cm.

4 Check your answer.

Top tip

- Make sure you measure the diameter through the centre of the circle.

Try this

1 Name the parts of the circle shown by the dashed lines. The centre of the circle is shown by a black dot.

a)

2 The diameter of a circle is 125 cm.

What is the radius? ☐ cm

b)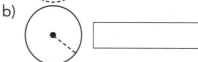

Coordinates

To achieve the higher score, you need to:

★ describe positions in all four quadrants on a 2-D coordinate grid

★ plot specified points and draw sides to complete a given polygon.

What you need to know

- There are four quadrants. In the first quadrant, the values of *x* and *y* are positive.
- The axes use negative numbers, so **coordinates** can use negative as well as positive numbers. You always read the coordinates in the same way; the first number is from the *x* (horizontal) **axis** and the second number is from the *y* (vertical) axis.

Let's practise

An isosceles triangle is drawn below. The coordinates of two vertices are shown. Write the missing coordinates of the third vertex.

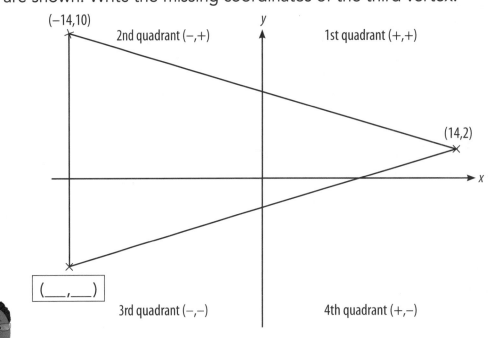

(−14,10)

2nd quadrant (−,+)

y

1st quadrant (+,+)

(14,2)

x

(___,___)

3rd quadrant (−,−)

4th quadrant (+,−)

1	Read the question and read it again. What is it asking?	Work out the coordinates for the third vertex of the isosceles triangle.
2	Two coordinates are given. Write known numbers on the axes using these coordinates.	The missing *x* coordinate is on the same line as another vertex: *x* = −14.
3	The name of the shape is given. Think about the properties of an isosceles triangle.	Two sides are equal. It has one line of symmetry that divides the shortest side of this triangle in half.

4 Use these facts to help find the missing coordinates.

To find the *y* coordinate, look at the
y coordinates of the other two vertices.
They are 10 and 2.
Moving down from *y* = 10 to *y* = 2 has moved
halfway down the shortest side of the triangle.
10 – 2 = 8, so half this side is 8 units. Move
down another 8 units to find the missing
y coordinate.
2 – 8 = –6

5 What is your answer?

(–14,–6)

6 Check that the coordinates line up.

Try this

1 Three coordinates of a rectangle are marked.

What are the coordinates of the missing vertex? (,)

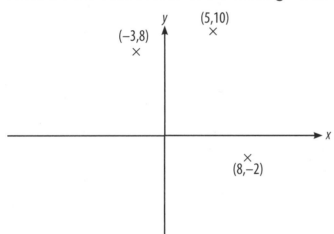

Top tip

- Always read along
  the *x* axis and then up/
  down the *y* axis. Write
  the *x* coordinate before
  the *y* coordinate: (*x, y*).

2 ABCD is a square.

On the grid, plot these points:

A (–3,4), B (4,4), C (4,–3).

Draw straight lines, using a ruler,
from A to B, to C.
Where is point D? (,)

Plot point D and complete the square.

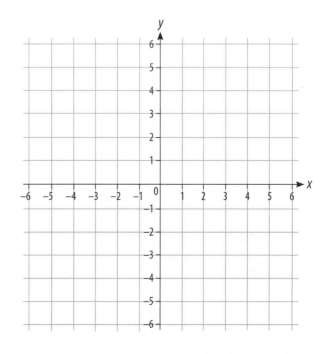

Translations

To achieve the higher score, you need to:

★ draw and translate simple shapes on the **coordinate plane**.

What you need to know

- A **translation** is an example of movements that change the position of a shape, but don't change its size or orientation.
- A translation can be described as a *slide*. It's a movement left or right, plus a movement up or down, but can be a combination of these.

Let's practise

This triangle is translated **5 right** and **3 down**. What are its new coordinates?

A (__,__)

B (__,__)

C (__,__)

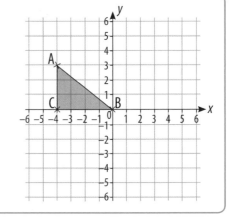

 1 Read the question and read it again. What is it asking?

Translate the triangle and work out its new coordinates.

2 Look at point A. Where will it be when the triangle is translated? What effect does the translation have on its coordinates?

Moving 5 to the right adds 5 to the x coordinate. Moving 3 down subtracts 3 from the y coordinate.
The new position for A is
$(-4 + 5, 3 - 3) = (1,0)$

3 Repeat for the other two points.

B goes to $(5,-3)$
C goes to $(1,-3)$

4 Plot the new points and draw the new triangle. Check it is the same size as the original triangle.

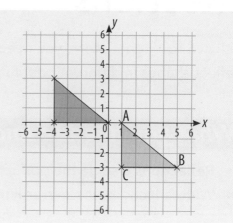

5 What is your answer?

A (1,0) B (5,–3) C (1,–3)

6 Check your calculations.

★ **Top tips**
- Make sure the translated **triangle** is identical to the original, just in a different place.
- Be careful to consider each coordinate in turn.

Try this

1 Label this diagram with the coordinates of the four vertices of this quadrilateral.

Then translate the quadrilateral **2 right** and **2 down**.

Write the new coordinates for each of the four vertices.

A (,)

B (,)

C (,)

D (,)

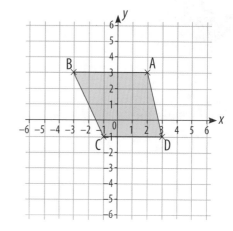

2 Shape A is translated.

The new position is shown by shape B. Describe the translation.

_____

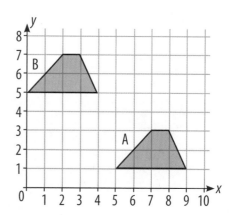

Reflections

To achieve the higher score, you need to:
★ reflect shapes in different orientations.

- **Reflections** change the position of a shape but don't change its size.
- With reflections, the **mirror line** shows where the shape is reflected.

Let's practise

Using the y axis as the mirror line, draw a reflection of the pentagon.
Write the new coordinates in the boxes below.

A (,)

B (,)

C (,)

D (,)

E (,)

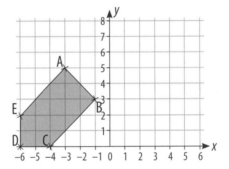

1 Read the question and read it again.
What is it asking?

Reflect the shape using the y axis as the mirror line.

2 Notice where each corner is positioned. In the reflected shape, each corner will be the same distance from the y axis but on the other side of it. When reflecting in the y axis, the x coordinate changes sign.

Corner A is at (−3,5). The reflected point A will be at (3,5).
Corner B is at (−1,3). Its reflection will be (1,3).
Corner C is at (−4,0). Its reflection will be (4,0).
Corner D is at (−6,0). Its reflection will be (6,0).
Corner E is at (−6,2). Its reflection will be (6,2).

3 Use a ruler to join the corners.
Check that the sides are all the same length as in the original shape.

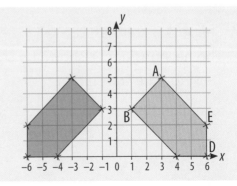

4 What is your answer?   (3,5) (1,3) (4,0) (6,0) (6,2)

5 Check your answer.

Try this

1 Shade two more squares to make shape 2 a reflection of shape 1 in the mirror line.

Shape 1   Shape 2

2 Reflect the shape in the mirror line.

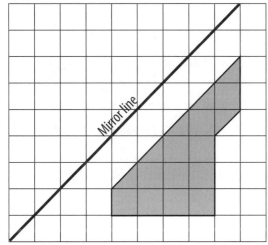

3 Reflecting in the y axis has the effect of changing the sign of the x coordinate.

How does reflection in the x axis affect the coordinates?

_____

_____

Top tip

- The reflection of a point is always on a line perpendicular to the mirror line, an equal distance the other side.

Tables

To achieve the higher score, you need to:

★ complete missing data in **tables** using problem-solving skills.

What you need to know

- **Tables** provide a way of presenting data, in rows and columns.

Let's practise

There are 380 children at Park School.
They can choose from four types of lunch.

Meal	Hot meal	Salad	Vegetarian	Sandwich	Total
Girls	72	15		23	175
Boys			26	20	
Total	105				

Complete the missing data in the table.

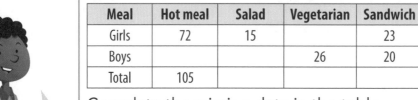

1 Read the question and read it again. What is it asking?

Fill in the missing numbers in the table.

2 Which can you complete with the data already given? Be systematic – look for columns or rows that have one missing number.

The number of boys having a hot meal = 105 − 72 = **33**. The number of girls having vegetarian lunch = 175 − 72 − 15 − 23 = **65**. The total having a sandwich is 23 + 20 = **43**.

3 Do your calculations help you to work out more numbers?

Yes, I can now work out the total for vegetarian. 65 + 26 = 91

4 Read the question again. Note that the total number of children is 380.

Now I can complete the rest of the numbers.

Meal	Hot meal	Salad	Vegetarian	Sandwich	Total
Girls	72	15	65	23	175
Boys	33	126	26	20	205
Total	105	141	91	43	380

5 Check your answers by adding up the rows and columns.

Try this

1 The table below shows the sizes and colours of T-shirts sold in a shop. Complete the table.

Size \ Colour	White	Blue	Red	Green	Yellow	Total
Small		6	3		9	43
Medium	12		7	11		
Large	5	14		3		45
Total		33	28	24	19	

Top tip

- Make sure all the rows and columns add up to the correct total.

Timetables

To achieve the higher score, you need to:
★ complete, read and interpret information in **timetables**.

What you need to know

- **Timetables** show events by time.

Let's practise

This is part of a train timetable.

London	09:00	09:30	10:00	10:30	10:50	11:00	11:30
Stevenage	09:50			10:49	11:22			12:20
Peterborough		10:28					12:02	12:35
Grantham	10:06		11:07			12:08	
Doncaster	10:35	11:06		12:12		12:31	13:02

Beth arrives at Stevenage station at quarter to 11 in the morning. She catches a train to Doncaster.

When will she arrive in Doncaster? ☐

1 Read the question and read it again. What is it asking?	**Find when Beth will arrive in Doncaster.**
2 Beth arrives at Stevenage station at quarter to 11. Change this time to 24-hour time so it is the same as the timetable.	**Quarter to 11 in the morning is 10:45.**
3 Use the row of times for trains from Stevenage and the row of times for Doncaster.	**Check times after 10:45 from Stevenage. 10:49 does not stop at Doncaster. 11:22 stops at Doncaster.**
4 Check the time the train arrives in Doncaster.	**The 11:22 train from Stevenage arrives in Doncaster at 12:12**
5 What is your answer?	**My answer is 12:12**
6 Check your answer.	

Try this

Use the same timetable to answer this question.

1 Ali is in London. He must arrive in Grantham by 12:00

Which is the latest train he should catch from London? ☐

Top tip

- A blank space on a timetable means the train or bus does not stop at that station.

Pictograms

To achieve the higher score, you need to:
★ interpret **pictograms** using a range of skills and strategies.

What you need to know

- **Pictograms** use images or symbols to represent numbers.
- The key on a pictogram shows what each symbol stands for.

Let's practise

The school cricket team scored exactly 200 runs in three games.

This pictogram shows the number of runs scored in each of the three games.

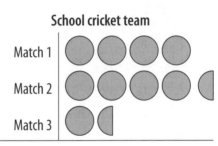

School cricket team

Match 1

Match 2

Match 3

 = 20 runs

What **fraction** of runs was scored in Match 3?

---

1 Read the question and read it again. What is it asking?

Work out the fraction of runs scored in Match 3.

2 Check the key. How much does each symbol represent?

One circle represents 20 runs.

3 To work out the fraction you need to know the total number of runs and the number of runs scored in Match 3.

200 runs were scored in total.
Match 3 shows $1\frac{1}{2}$ circles. This is one group of 20 runs and half a group, which will be 10 runs. So, 30 runs in total.

4 What is 30 runs as a fraction of the total number of runs?

$\frac{30}{200}$, which simplifies to $\frac{3}{20}$, so $\frac{3}{20}$ of the runs were scored in Match 3.

5 What is the answer?

$\frac{3}{20}$

6 Check your answer.

Try this

1 Using the pictogram on page 56, find out the **ratio** of runs scored in Match 1, Match 2 and Match 3 [: :]

2 This pictogram shows the number of spectators, rounded to the nearest 5,000, at five football matches.

KEY: ♂ = 5,000 spectators

Football attendances	
West Ham v Crystal Palace	♂♂♂♂♂♂♂
Burnley v Swansea	♂♂♂
Newcastle v Aston Villa	♂♂♂♂♂♂♂♂♂♂
Stoke v Hull	♂♂♂♂♂
West Brom v Southampton	♂♂♂♂♂

a) What **fraction** of all the spectators attended the Newcastle v Aston Villa match? [ ]

b) What **percentage** of spectators attended the Burnley v Swansea match? [ ]

c) What was the **ratio** of spectators at the Stoke v Hull game to spectators at the other four games? [ ]

3 A town has four car parks. The number of spaces in each car park is shown by a symbol.

KEY: 🚗 = 25 spaces

Spaces in each car park	
High Street Car Park	🚗🚗🚗🚗🚗🚗🚗
Market Car Park	🚗🚗🚗🚗🚗🚗🚗🚗🚗
Riverside Car Park	🚗🚗🚗🚗🚗
Wood Street Car Park	🚗🚗🚗🚗🚗🚗🚗🚗

a) How many spaces are there in the High Street Car Park? [ ]

b) How many more spaces are there in the Market Car Park than in the Riverside Car Park?

[ ]

c) Wood Street Car Park charges £1.50 an hour.
   When the car park is full, how much money is taken in an hour? £[ ]

Top tip

• Always read the pictogram key carefully.

Bar charts

To achieve the higher score, you need to:

★ interpret **bar charts** using a range of skills and strategies.

What you need to know

- Compound **bar charts** show grouped data in stacked (compound) columns.

Let's practise

This bar chart shows information about the membership of a gym club.

What **fraction** of the **total** membership is over 30 years of age?

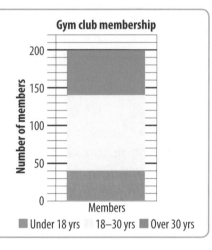

Gym club membership

Number of members

200

150

100

50

0

Members

■ Under 18 yrs ☐ 18–30 yrs ■ Over 30 yrs

1 Read the question and read it again. What is it asking?

I need to find the number of members over 30 years of age and use this to find the fraction of the total membership.

2 Study the bar and find which bit means *over 30*.

The red shading goes from 140 to 200 members, so 60 members in the club are over 30.

3 Find the fraction of all members over 30 years of age as a fraction of the total membership.

60 members are over 30 years.
Total membership = 200
$\frac{60}{200}$ simplifies to $\frac{3}{10}$.

4 What is your answer?

$\frac{3}{10}$ of the members are over 30 years of age.

5 Check your answer.

Top tip

- Label separate segments within each bar to help you answer the questions.

Try this

Use the bar chart above to answer these questions.

1 What is the **ratio** of members over 30 years of age to the rest of the membership? ☐ : ☐

2 What **fraction** of the whole membership is under 18 years of age? ☐

Pie charts

To achieve the higher score, you need to:
* ★ interpret and construct **pie charts**
* ★ use pie charts to solve problems.

What you need to know

- A **pie chart** shows the proportion of each category of data. Pie charts are good for showing data as **percentages** or fractions of the whole.
- The angle at the centre of the pie chart is 360°.

Let's practise

Kamal conducts a survey of 120 pupils to find out how they spend their pocket money.

This table shows the data Kamal collected.

Draw a pie chart for this data.

What they spend their money on	Number of pupils
hobbies	45
clothes	35
music	40

1 Read the question and read it again. What is it asking?

Draw a pie chart for the data in the table.

2 Find the total number of pupils. Then divide 360° by the total number of pupils to find out what angle to use to represent 1 pupil.

$45 + 35 + 40 = 120$
$360° ÷ 120 = 3°$

3 Multiply each option by this angle (3°) to work out the size of each slice of the pie chart.

What they spend their money on	Number of pupils	Angle for the slice
hobbies	45	$45 × 3 = 135°$
clothes	35	$35 × 3 = 105°$
music	40	$40 × 3 = 120°$

4 Draw a circle, and use a protractor to mark off each slice. Label each slice.

What pupils spend their money on

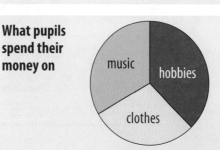

5 Check your answer.

Try this

1 Another 60 children are added to the survey. What size angle will be needed to represent one pupil now? ☐ °

Top tip

- Make sure all the slices add up to 360 degrees.

Line graphs

To achieve the higher score, you need to:
★ interpret and construct line graphs and use these to solve problems.

★ What you need to know

- **Line graphs** are used to present **continuous data**.
  This means that the number can take any value (e.g. a boy's mass such as 32.75 kg could include fractions or decimals of a kilogram).

★ Let's practise

This graph is used to change miles and kilometres.

How many miles is 40 kilometres? ☐

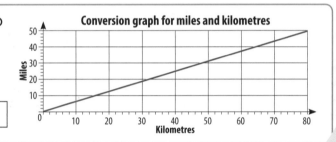

Conversion graph for miles and kilometres

Miles / Kilometres

1 Read the question and read it again. What is it asking?

Use the graph to change 40 kilometres into miles.

2 Check the divisions on the graph.

The major divisions increase by 10 each time. Five small divisions represent 10 units on both axes, so each small division is 2 units.

3 To complete the conversion, find 40 kilometres on the *x* (horizontal) axis.
Then move vertically up the line until you reach the conversion line. Move horizontally to the *y* axis.

The conversion line shows that 40 kilometres equals 25 miles.

4 What is your answer?

40 kilometres = 25 miles

5 Check your answer by reading the graph from miles to kilometres.

★ Try this

Answer these questions using the same graph.

1 Change 28 kilometres into miles. ☐ miles

2 Use the graph to convert 30 miles into kilometres. ☐ km

★ Top tip

- Work out the value of unmarked divisions by using the spaces between known values.

Averages

To achieve the higher score, you need to:
★ calculate the **mean** as an **average** for simple sets of discrete data in different contexts.

What you need to know

- An **average** is a value that represents a large set of data.
- One example of an average is the **mean**. To find the mean, add all the amounts and divide the total by the number of amounts.

Let's practise

> The mean of four numbers is 23.2
> A fifth number is added and the new mean is 19.6
> What is the **fifth** number? ☐

1	Read the question and read it again. What is it asking?	Find the fifth number of a set of numbers by using the mean.
2	If the mean of four numbers is 23.2, work out the total of the four numbers.	$23.2 \times 4 = 92.8$
3	If the mean of five numbers is 19.6, work out the total of the five numbers.	$19.6 \times 5 = 98$
4	The total has increased. The increase must be the fifth number.	$98 - 92.8 = 5.2$
5	What is your answer?	5.2
6	Check your answer.	$23.2 \times 4 + 5.2$ divided by $5 = 19.6$

Try this

1 The mean of six numbers is 34.8
   What is the **total** of the six numbers?
   ☐

2 The mean of three numbers is 21.625
   Two of the numbers are 15.999 and 32.15
   What is the third number? ☐

Top tip

- The mean value is **not** one of the data items. The mean is calculated as the sum of all the values divided by the number of values.

3 The mean of four numbers is 5.2
   One of the numbers is removed and the new mean is 2.9
   What number was removed? ☐

Glossary

Algebra *See Formula. See Variable.* A system that uses unknowns, such as letters or symbols, in place of numbers.

Angle A turn measured in degrees. A full turn is 360 degrees (360°).

Axis A line that is used on a graph to count from or build blocks and bars. The horizontal axis goes from left to right and is called the x axis. The vertical axis goes up and down and is called the y axis. The plural is *axes*.

Capacity *See Volume.* The volume of a liquid held in a container; measured in litres or millilitres. It can also be measured with cubes of the same size, such as cubic centimetres (cm³) or cubic metres (m³). 1 ml = 1 cm³. 1 l = 1,000 cm³.

Circumference The outside edge or perimeter of a circle.

Common factor A factor that is common to two or more numbers, e.g. 1 and 3 are the common factors of 18 and 21.

Common multiple A multiple that is common to two or more numbers, e.g. the common multiples of 4 and 5 are 20, 40 and any multiples of those numbers.

Composite shape A shape made from two or more other shapes.
This is a composite shape:

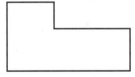

It is made from two rectangles:

Continuous data Numbers or data that can take any value, e.g. the mass of an object is not limited to whole numbers since fractions and decimals can be used to give a mass.

Correspondence Two sides on two shapes correspond if they are in a similar position, e.g. side A is in a corresponding position to side B.

 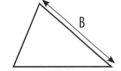

Cube number The result of multiplying a number by itself twice, e.g. 5 × 5 × 5 = 125 so 125 is a cube number. The abbreviated way to write 5 × 5 × 5 is to write 5^3 (5 cubed).

Decimal place The position of a digit in a decimal number. The places are numbered from left to right following the decimal point, e.g. in the number 35.627:
- 6 is in the first decimal place
- 2 is in the second decimal place
- 7 is in the third decimal place.

Diameter The distance from a point on the circumference of a circle to another point on the circumference through the centre of the circle.

Equation When two terms or expressions are equal to each other, e.g. 5 + 4 = 3 × 3 or 5n + 10 = 45.

Formula *See Algebra. See Variable.* A set of mathematical instructions that uses numbers to calculate the answer to a problem, e.g. a formula to find the perimeter of a rectangle is $P = 2(l + w)$ where l is the length of the rectangle and w is the width of the rectangle. The plural is *formulae*.

Imperial unit An older set of units used for measuring, including inch, foot, yard, mile (length); ounce, pound, stone (mass); pint, gallon (capacity).

Linear number sequence Numbers that follow a pattern according to a rule (or a formula). The rule can be described by an expression using n.

Mass Similar to the weight of an object, but mass is constant and the weight of an object can vary. Grams (g) and kilograms (kg) are two of the more common measures for mass.

nth term rule A way of describing a sequence of numbers and calculating any term (number) in the sequence, e.g. an adding 4 sequence is really writing down numbers that have a relationship to the four times table:
7   11   15   19   23
So, the sequence can be described as 4n because of this relationship. But this sequence is the multiples of 4 plus 3, so the nth term rule is 4n + 3. The 22nd number in the sequence can be found by substituting 22 for n: 4 × 22 + 3 = 91, so 91 is the 22nd number in this sequence.

Order of operations In calculations that involve more than one calculation, there is a special order for completing the operations. BIDMAS and BODMAS are ways of remembering this order.
BIDMAS stands for Brackets, Indices, Division, Multiplication, Addition, Subtraction.
BODMAS stands for Brackets, Order, Division, Multiplication, Addition, Subtraction.
This is the order in which calculations should be completed when more than one operation is given in a calculation.
In calculations with only multiplication and division, work from left to right, e.g.
$5 \times 2 \div 5 = 10 \div 5 = 6$.
In calculations with only addition and subtraction, work from left to right, e.g.
$10 - 6 + 4 = 4 + 4 = 8$.

Pie chart A circular graph that shows the proportion of different amounts by using the sectors of a circle.

Prime factor *See Prime number.* A group of prime numbers that multiply to make another number, e.g. the prime factors of 30 are 2; 3; and 5. 2; 3; and 5 are all prime numbers.
$2 \times 3 \times 5 = 30$

Prime number A number that only has two factors: itself and 1, e.g. 23 is a prime number as its only factors are 23 and 1. Note that 1 is not a prime number as it only has one factor, which is 1. Remember that 2 is the only prime number that is an even number.

Proper fraction A fraction that is less than 1. A proper fraction has a numerator that is less than the denominator, e.g. $\frac{3}{5}$, $\frac{9}{10}$ and $\frac{1}{2}$ are all proper fractions.

Proportion *See Ratio.* Keeping the relative sizes of two or more numbers or amounts equal, e.g. 2 for every 5 is in proportion with 4 for every 10 and 6 for every 15. Proportion keeps ratios equal, e.g. $2:5 = 4:10 = 6:15$.

Radius The distance from the centre of a circle to a point on the circumference.

Ratio *See Proportion.* Compares the sizes of two or more numbers or amounts. Ratios are usually written with a colon and the word 'to' is said in place of the colon, e.g. $3:1:5$ (say 'three to one to five').

Reflection A reflection of a shape flips the whole shape in a mirror line or a line of reflection.

Roman numerals The system of recording numbers used by the Romans. They used letters to write the value of each digit individually. The letters used were: I = 1; V = 5; X = 10; L = 50; C = 100; D = 500; M = 1000.

Rule A description of how a sequence increases or decreases.
13   21   29   37   45
This sequence increases by 8 each time, so the rule would be 'add eight' or '+8'.
60   57   54   51   48
This sequence decreases by 3 each time, so the rule would be 'subtract/minus three' or '–3'.

Scale factor The number used to increase or decrease a value or values, e.g. if the length of a line has increased from 5 cm to 20 cm, the scale factor is 4 because $5 \times 4 = 20$.

Symbol A letter or a drawing that is used in place of a number or a word.

Timetable A table that indicates the time of an event. Railway timetables can show the arrival and departure times of trains or a class timetable can show the time a lesson starts and ends.

Unequal sharing *See Ratio.* A number or an amount that has been divided so each division is unequal. This can occur in different ways using fractions, percentages or ratios.

Unit fraction A fraction where the numerator is 1, e.g. $\frac{1}{2}$, $\frac{1}{3}$, $\frac{1}{4}$.

Variable *See Algebra. See Formula.* A number or amount that may change. In the formula to find the perimeter of a rectangle, $P = 2(l + w)$, the length and width will change depending on the size of the rectangle. These are the variables.

Volume *See Capacity.* The amount of space taken up by a 3-D shape or object. Volume can refer to solids or empty spaces and is measured using cubes of the same size, such as centimetre cubes (cm^3) or metre cubes (m^3).

Answers

Number and place value

Place value (page 8)
1 a) $\frac{3}{10}$ or three-tenths

  b) $\frac{3}{100}$ or three-hundredths

  c) $\frac{3}{1,000}$ or three-thousandths

  d) 3,000 or three thousand

  e) 30 or three tens

Roman numerals (page 9)
1 a) CCLXV
  b) CMXCIX
  c) MM
  d) MMXV
2 a) 2,014
  b) 286
  c) 89
  d) 124

Number – Addition, subtraction, multiplication and division

Addition and subtraction (page 10)
1 718

Squares and cubes (page 11)
1 225; 256; 289 (accept 15^2; 16^2; 17^2; do not accept 15; 16; 17)
2 7

Common multiples (page 12)
1 24; 48; 72; or any other multiple of 24
2 9
3 120 and 144

Common factors (page 13)
1 a) 120; 168; or 192 and every subsequent multiple of 24 to 984
  b) 1,008; 1,032; 1,056 and every subsequent multiple of 24 to 9,984
2 Explanation needs to state that every number has 1, an odd number, as a factor.

Prime numbers and prime factors (page 14)
1 a) $2 \times 3 \times 3$ (accept 2×3^2)
  b) $2 \times 2 \times 7$ (accept $2^2 \times 7$)
2 79; 83; 89; 97

Multiplying and dividing by 10, 100 and 1,000 (page 15)
1 a) 920.7 b) 1,000 c) 306 d) 100
2 Explanation should show that the answer should be 103 **or** that Zoe has put the numbers in the wrong order.

Long multiplication (page 16)
1 111,456 km (8 journeys each month = 1161 × 96)

Long division (page 17)
1 100 boxes

Correspondence (page 18)
1 216

Order of operations (page 19)
1 17.5
2 $10 \times (10 \times 10 - 10) = 900$

Number – Fractions, decimals and percentages

Ordering fractions (page 20)
1 a) $\frac{7}{10}$ $\frac{2}{3}$ $\frac{3}{5}$ (accept $\frac{21}{30}$ $\frac{20}{30}$ $\frac{18}{30}$)

  b) $4\frac{2}{3}$ $4\frac{5}{8}$ $4\frac{7}{12}$ $3\frac{5}{6}$

  (accept $4\frac{16}{24}$ $4\frac{15}{24}$ $4\frac{14}{24}$ $3\frac{20}{24}$)

Adding and subtracting fractions (page 21)
1 $8\frac{1}{24}$
2 $1\frac{9}{20}$
3 $1\frac{17}{30}$
4 $9\frac{1}{24}$

Multiplying fractions (page 22)
1 $\frac{9}{32}$
2 $\frac{3}{14}$ (accept $\frac{6}{28}$)
3 a) $1\frac{7}{8}$ b) 9

Dividing fractions (page 23)
1 a) $\frac{1}{12}$ b) $\frac{1}{32}$
2 a) 5 b) 3

Changing fractions to decimals (page 24)
1 3.5
2 0.125
3 4.625

Rounding decimals (page 25)
1 a) 25.7 b) 25.73
2 37.638 and 37.58 ticked

Adding and subtracting decimals (page 26)
1 101.436
2 37.845
3 76.267
4 7.745

Multiplying decimals (page 27)
1 0.35
2 4.8
3 0.36
4 8.1
5 7
6 0.09
7 6
8 4

Percentages (page 28)
1 £15.60
2 £80

Ratio and proportion

Ratio (page 29)
1 297
2 69

Proportion (page 30)
1 12.5
2 a) 270 b) $\frac{5}{9}$ c) 4:5

Scaling problems (page 31)
1 13.5
2 27, 7.5 cm

Unequal sharing (page 32)
1 £630
£240
£130
2 $\frac{1}{5}$

Algebra

Algebra (page 33)
1 a) $2y - 7 = 51$ b) 29
2 a) $(y \div 2) \times 3 = 27$ b) 18

Using formulae (page 34)
1 £133

Solving equations (page 35)
1 △ = 8   ◇ = 5
2 $R = 20$, $T = 4$

Linear number sequences (page 36)
1 $5n + 1$
2 8; 11; 14; 17; 20

Measurement

Measures (page 37)
1 125 ml

Converting metric units (page 38)
1 150 ml tub costs 30p/ml, 250 ml tub costs 24p/ml,
600 ml tub costs 25p/ml
So the 250 ml tub is the best value and should be ticked.
2 £5.85

Metric units and imperial measures (page 39)
1 60 cm +/− 5 cm
2 6.3 kg +/− 0.5 kg
3 14p +/− 1p

Perimeter and area (page 41)
1 Rectangle drawn with an area of 21 cm².
2 Taking the length to be the longest side: Rectangle with area of
24 cm² = length of 8 cm and **width of 3 cm**.
Rectangle with perimeter of 24 cm² = length of 8 cm and **width of 4 cm**.
3 a) 184   b) 1,401

Area of parallelograms and triangles (page 43)
1 560
2 100
3 135

Volume (pages 44–45)
1 One measurement must be 15 cm. The other lengths could be
30 cm and 1 cm, 15 cm and 2 cm, 10 cm and 3 cm, 6 cm and 5 cm
(accept lengths using fractions or decimal lengths that multiply
to 30 cm², e.g. 12 cm × 2.5 cm).
2 Accept any three numbers that multiply to 300 cm³,
e.g. 300 × 1 × 1 or 10 × 6 × 5.
3 729
4 Accept any three measurements that multiply to 64,
e.g. 64 × 1 × 1, 32 × 2 × 1, 8 × 4 × 2
5 50

Geometry – Properties of shapes

Angles and degrees (page 46)
1 $x = 57$, $y = 106$

Circles (page 47)
1 a) circumference   b) radius
2 62.5

Geometry – Position and direction

Coordinates (page 49)
1 (0,−4)
2 (−3,−3)

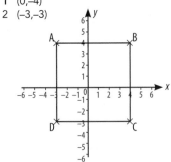

Translations (page 51)
1 A (4,1)   B (−1,1)   C (1,−3)   D (5,−3)
2 Shape A is translated to Shape B left 5, up 4

Reflections (page 53)
1

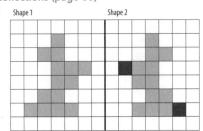

2

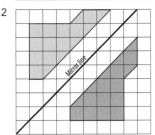

3 The sign of the y coordinates will change, e.g. coordinate (3,5)
reflected in the x axis will be (3,−5).

Statistics

Tables (page 54)
1

Colour / Size	White	Blue	Red	Green	Yellow	Total
Small	15	6	3	10	9	43
Medium	12	13	7	11	5	48
Large	5	14	18	3	5	45
Total	32	33	28	24	19	136

Timetables (page 55)
1 10:00

Pictograms (page 57)
1 8:9:3 (accept 80:90:30)
2 a) $\frac{1}{3}$ (accept $\frac{10}{30}$)   b) 10%   c) 1:5 (accept 5:25)
3 a) 175 spaces   b) 100 spaces   c) £300

Bar charts (page 58)
1 3:7 (accept 6:14)
2 $\frac{1}{5}$ (accept $\frac{4}{20}$)

Pie charts (page 59)
1 2

Line graphs (page 60)
1 18 +/− 2
2 48 +/− 2

Averages (page 61)
1 208.8
2 16.726
3 12.1